Key to
Metric Measurement®

Measuring Length and Perimeter Using Metric Units

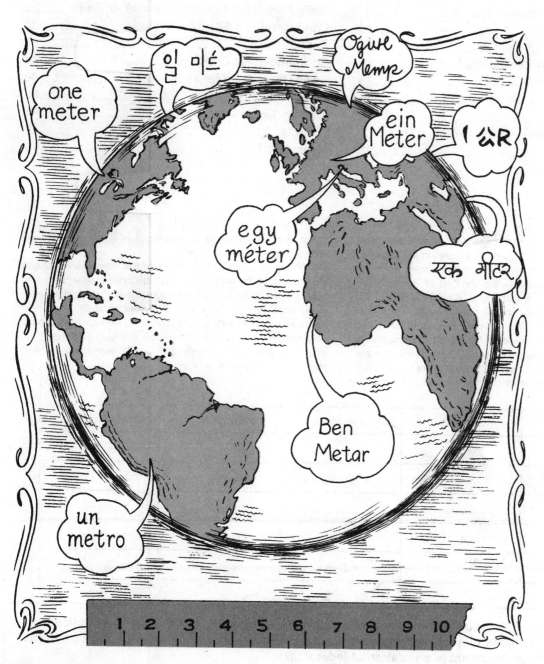

Betsy Franco

Name _____   Class _____

CONTENTS

Lend a Hand 1
Choosing a Measuring Tool 2
Team Project #1: Making a Trundle Wheel 3
Estimating Measurements 4
Changing Units of Length 5
Finding the Better Buy 6
Talking About Length 7
Comparing Heights 8
Metric Words 9
Planetary Jumping 12
Review: Adding, Subtracting, and
   Multiplying Decimals 13
Olympic Records 14
Review: Measuring in Centimeters
   and Millimeters 15
Renting Yourself Some Ski Equipment 16
Metric Tools 17
Olympic Training Center—Using Meters
   and Kilometers 18

More About Kilometers 19
A Distance Chart for Mexican Cities 20
Odometer Readings 21
Estimation or Careful Measurement? 22
Introduction to Perimeter 23
Measuring Perimeters 24
Exploration of Perimeters 25
Calculating Perimeters 26
Missing Measurements 27
Team Project #2: The Secret of Circumference 28
Team Project #3: Circumference All Around 30
Can You Read a Floorplan? 32
Reductions 33
Scale and Scale Drawings 36
Scale Drawing of Maria's House 38
Scale on a Map 40
Practice Test 41
Projects 43
Measuring Tools 45

Numbers in Many Languages

Do you know how the word *meter* originated? *Meter* comes from the Greek word *métron*, which means "measure." In this book you will see many of the different units of the metric system. Each unit of measurement has a prefix as the first part of its name—for example, *centimeter* has the prefix *centi. Centi* is related to the Greek word for "hundred" and means one-hundredth. *Deci* (one-tenth) and *milli* (one-thousandth) are other prefixes in the metric system that are related to Greek words. The chart below shows numbers in different languages. Because these languages are closely related to each other, many of the words are similar. Also, many of these numbers are related to the prefixes for metric units.

Which of these languages can you speak or understand? Do you see how the words are related?

You might conclude from the chart that these five languages share some of the same roots. Perhaps those languages are all related to the same "parent" language. However, many languages in the world do not resemble these languages at all. If you hear words for numbers in these languages, you will notice they sound very different. Despite these differences, nearly all countries in the world use the metric system, and people from many different countries say "meter" in more or less the same way.

Latin	Spanish	French	German	English (U.S.)
unus	uno	un	eins	one
decem	diez	dix	zehn	ten
centum	ciento	cent	hundert	hundred
mille	mil	mille	tausend	thousand
	millón	million	million	million

IMPORTANT NOTICE: This book is sold as a student workbook and is *not* to be used as a duplicating master. No part of this book may be reproduced in any form without the prior written permission of the publisher. Copyright infringement is a violation of federal law.

Copyright © 2000 by Key Curriculum Press. All rights reserved.
® *Key to Fractions, Key to Decimals, Key to Percents, Key to Algebra, Key to Geometry, Key to Measurement,* and
Key to Metric Measurement are registered trademarks of Key Curriculum Press.
Published by Key Curriculum Press, 1150 65th Street, Emeryville, CA 94608
Printed in the United States of America   10 9 8 7 6 5 4 3   03 02 01 00 99   ISBN 1-55953-326-9

Lend a Hand

Spread your fingers as far apart as you can and trace your hand on this page. Draw a line segment showing each distance below. Measure each line segment to the nearest centimeter.

hand span _____

length of longest finger _____

length of little finger _____

width of index finger _____

width of wrist _____

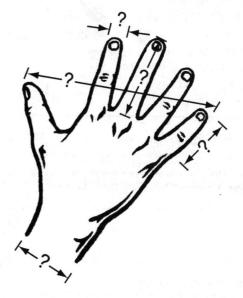

Did you remember to write the unit of measure with each answer?   ☐ Yes   ☐ No

©2000 by Key Curriculum Press
Do not duplicate without permission.

Choosing a Measuring Tool

Different tools work best for different jobs.

1. measuring tape	2. tape measure	3. trundle wheel

4. ruler	**5. car odometer**	A trundle wheel is used to measure distances too long for a measuring tape. You count how many times the wheel goes around and multiply by the distance around the wheel.
	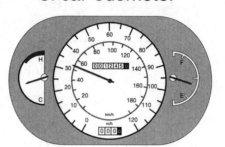	

Write the number of the best measuring tool for each situation.
You may choose more than one for each answer.

_____ Barbara is measuring the width of a garden plot in her backyard. She plans to plant a row of tomatoes and a row of cucumbers.

_____ Raymond's dad is measuring a route in their neighborhood. He and Raymond plan to jog 5 kilometers together once a week.

_____ Kwaku is measuring her broken necklace chain before she buys a new one to replace it.

_____ Gordon is measuring the length of his favorite pants before going to the store to buy some jeans.

_____ Barney is measuring the length and width of the park to determine its area. He needs to know the area before he plants new grass.

_____ Li is measuring around her head to order a hat over the telephone.

_____ Rosita is marking the soccer fields at the middle school for a soccer tournament.

©2000 by Key Curriculum Press
Do not duplicate without permission.

Team Project #1: Making a Trundle Wheel

Have you ever used a trundle wheel to measure a distance? To make your trundle wheel, you'll need masking tape or duct tape, an index card, a bicycle, a piece of chalk, and measuring tape. See page 45 if you need a measuring tape.

Tape half of an index card onto one of the spokes of the bicycle's front wheel so that it sticks out on the side, as shown at the right.

Measure the distance around the wheel to the nearest centimeter. Mark this distance on the wheel in equal segments, like on a ruler. You measure the distance around the wheel in centimeters so that when you measure long distances in meters, your measures will be more accurate. Remember, to go from centimeters to meters, you must divide by 100. For example, 205 centimeters is 2.05 meters.

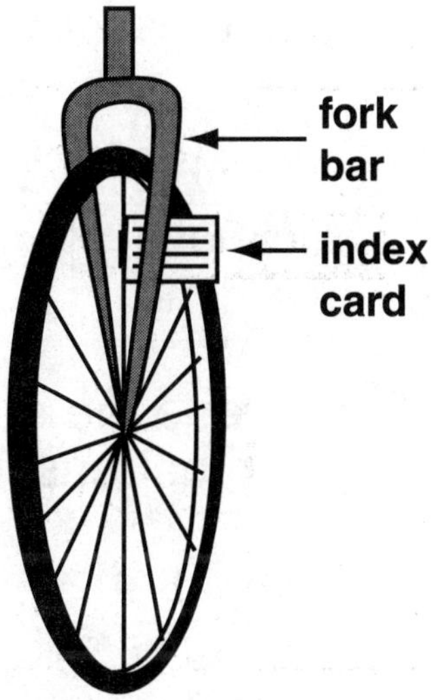

Rotate the wheel until the card touches the fork bar. To measure a distance, move the bike forward and count the number of times the card hits the fork bar. Multiply the number of hits by the distance around the wheel. Use the marks you made on the wheel to measure partial rotations.

fork bar

index card

Measure the distances given below. Give your answers in meters.

- the length of the longest wall of your house or apartment _____

- the distance from your room to the kitchen _____

- the distance from the front door to the nearest stop sign _____

- the distance around your house or apartment building _____

- the length of a block in your neighborhood _____

Now find two other distances to measure using your trundle wheel. Describe each one below.

_____

_____

_____

©2000 by Key Curriculum Press
Do not duplicate without permission.

Estimating Measurements

Before you make a measurement, it is often helpful to make a reasonable guess of that measurement. A reasonable guess is called an **estimate**. For instance, perhaps you can tell that the pencil shown below is longer than 5 centimeters and shorter than 10 centimeters. An estimate might be 8 centimeters.

I think this is a reasonable estimate. ☐ Yes ☐ No

The objects described below provide rough guides to help you visualize some units of measure. Think about these objects when you estimate length.

1 millimeter (mm)	1 centimeter (cm)	1 decimeter (dm)	1 meter (m)	1 kilometer (km)
thickness of a dime	width of a pinkie	width of an adult's hand	height of a doorhandle	distance of 5 city blocks

Terry made the estimates below. Make an ✗ through each one that seems too far off to be reasonable.

The height of the ceiling is 3.5 meters.

The height of a refrigerator is 180 centimeters.

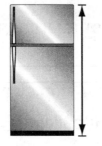

The length of a needle is 3 millimeters.

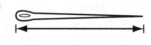

The length of a school yard is 2 kilometers.

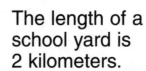

The height of a countertop is 4 meters.

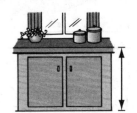

Make your own estimate for each length below. Then measure with a ruler or a meter stick to see if your estimate was close.

The length of your pinkie fingernail in millimeters is _____ millimeters.

Was your estimate close? _____

The length of the room is _____ meters.

Was your estimate close? _____

©2000 by Key Curriculum Press
Do not duplicate without permission.

Changing Units of Length

The millimeter, centimeter, and meter are metric units of length. You can use any of these units to measure the same length.

Use a tape measure, ruler, or meterstick to fill in the following three blanks.

1 cm = _____ mm   1 m = _____ cm   1 m = _____ mm

Multiply to change larger units, like centimeters, into smaller units, like millimeters. Fill in the blanks.

2 cm = __**20**__ mm

> 1 cm = 10 mm, so multiply the number of centimeters by 10.
> 2 x 10 = 20

6 cm = _____ mm   75 cm = _____ mm

17 m = _____ cm   12 m = _____ cm

4 m = _____ mm   2 m = _____ mm

8 cm = _____ mm   24 m = _____ cm

Divide to change smaller units, like millimeters, into larger units, like meters. Fill in the blanks.

300 cm = __**3**__ m

> 100 cm = 1 m, so divide the number of centimeters by 100.
> 300 ÷ 100 = 3

400 cm = _____ m   50 cm = _____ m

80 mm = _____ cm   170 mm = _____ cm

3000 mm = _____ m   2000 mm = _____ m

7000 cm = _____ m   240 mm = _____ cm

Look at the units in each problem. Then decide whether to multiply or divide. Fill in the blanks with the correct numbers.

26 cm = _____ mm   800 cm = _____ m   11 m = _____ cm

3 m = _____ mm   320 mm = _____ cm   2400 cm = _____ m

©2000 by Key Curriculum Press
Do not duplicate without permission.

Finding the Better Buy

Both items in each pair shown below are the same price. Change the larger unit into the smaller unit by multiplying. Then circle the item that is the better buy.

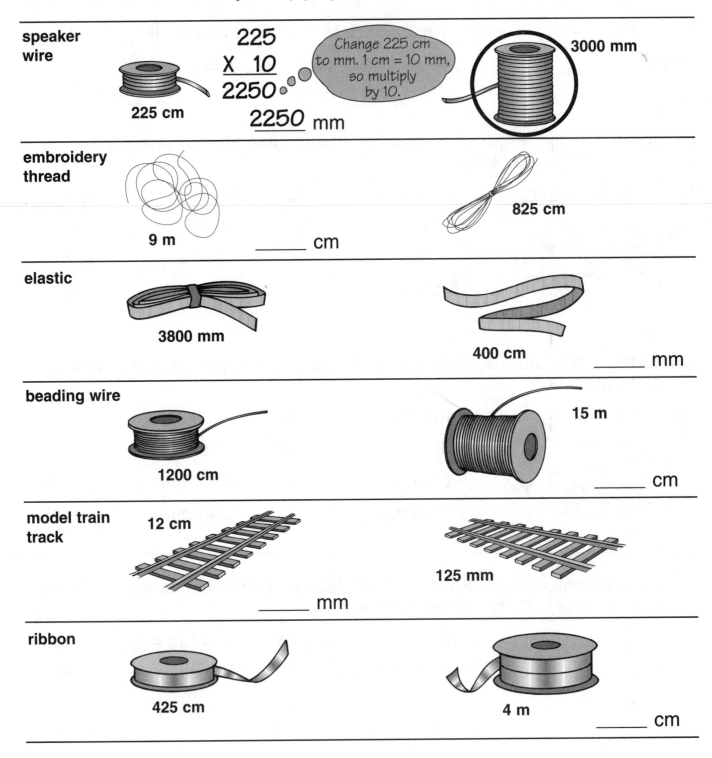

speaker wire

225 cm

$$225 \times 10 \over 2250$$

Change 225 cm to mm. 1 cm = 10 mm, so multiply by 10.

<u>2250</u> mm

3000 mm

embroidery thread

9 m

_____ cm

825 cm

elastic

3800 mm

400 cm

_____ mm

beading wire

1200 cm

15 m

_____ cm

model train track

12 cm

125 mm

_____ mm

ribbon

425 cm

4 m

_____ cm

©2000 by Key Curriculum Press
Do not duplicate without permission.

Talking About Length

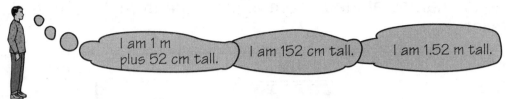

Make each measurement below using a meterstick or measuring tape. Then fill in the blanks.

I am _____ m plus _____ cm tall. I am _____ cm tall.

I am _____ m tall.

The door is _____ m plus _____ cm high. The door is _____ cm high.

The door is _____ m high.

The ceiling is _____ m plus _____ cm from the floor.

The ceiling is _____ cm from the floor. The ceiling is _____ m from the floor.

Sometimes you may want to use only centimeters or only meters to measure length.

Start with 4 m + 32 cm. 1 m = 100 cm, so multiply 4 by 100 and add the extra centimeters. 4 × 100 = 400. 400 + 32 = 432 cm.

4 m + 32 cm = 432 cm = 4.32 m

7 m + 26 cm = _____ cm = _____ m

8 m + 15 cm = _____ cm = _____ m

5 m + 84 cm = _____ cm = _____ m

11 m + 32 cm = _____ cm = _____ m

Start with 432 cm. 100 cm = 1 m, so divide 432 by 100 by moving the decimal point. 432 ÷ 100 = 4.32 m.

Sometimes you may want to change centimeters or meters to meters and centimeters.

100 cm = 1 m, so divide 368 by 100. Any number left over still represents centimeters. 3 m plus 68 cm are left over.

The whole number 4 is the number of meters. The decimal 0.87 means $\frac{87}{100}$ of a meter. 1 cm = $\frac{1}{100}$ m, so $\frac{87}{100}$ of a meter is 87 cm.

368 cm = 3 m + 68 cm

584 cm = _____ m + _____ cm

913 cm = _____ m + _____ cm

307 cm = _____ m + _____ cm

4.87 m = 4 m + 87 cm

2.95 m = _____ m + _____ cm

6.28 m = _____ m + _____ cm

12.54 m = _____ m + _____ cm

©2000 by Key Curriculum Press
Do not duplicate without permission.

Comparing Heights

The heights of nine athletes are given below. Write all measurements in centimeters. List the names at the bottom of the page from the shortest athlete to the tallest.

Azim	Alla	Amy

1.93 m = _____ cm	1.52 m = _____ cm	1.63 m = _____ cm

Bert	Bob	Beau

1.83 m = _____ cm	1.75 m = _____ cm	2.03 m = _____ cm

Candice	Carmen	Cody

1.71 m = _____ cm	160 cm = _____ cm	1.22 m = _____ cm

1. _____

2. _____

3. _____

4. _____

5. _____

6. _____

7. _____

8. _____

9. _____

©2000 by Key Curriculum Press
Do not duplicate without permission.

Metric Words

List three words that use each metric prefix. Use a dictionary if you need it.

centi: _____ _____ _____

deci: _____ _____ _____

milli: _____ _____ _____

kilo: _____ _____ _____

Write the meaning of each word below. Use a dictionary if you need it.

millennium: _____

century: _____

centigrade: _____

hectare: _____

decibel: _____

decade: _____

hectometer: _____

decameter: _____

kilowatt: _____

decapod: _____

©2000 by Key Curriculum Press
Do not duplicate without permission.

Now look again at the names of metric units of length. The chart below shows how to change all the metric measurements of length into meters. Fill in the blanks and look for a pattern.

There are <u>1000 meters</u> in a kilometer.	To change kilometers to meters, <u>multiply by 1000</u>.
There are <u>100 meters</u> in a hectometer.	To change hectometers to meters, <u>multiply by 100</u>.
There are <u>10 meters</u> in a decameter.	To change decameters to meters, <u>multiply by</u> _____.
There is <u>1 meter</u> in a meter.	To change meters to meters, _____.

There is <u>0.1 meter</u> in a decimeter.	To change decimeters to meters, <u>divide by 10</u>.
There is <u>0.01 meter</u> in a centimeter.	To change centimeters to meters, <u>divide by</u> _____.
There is <u>0.001 meter</u> in a millimeter.	To change millimeters to meters, _____.

Fill in the blanks.

How many meters in a ...	To change to meters,	Example
kilometer? <u>1000</u>	<u>multiply by 1000.</u>	45 km = <u>45 000</u> m
hectometer? _____	_____	45 hm = _____ m
decameter? _____	_____	45 dam = _____ m
meter? _____	_____	45 m = _____ m
decimeter? _____	_____	45 dm = _____ m
centimeter? _____	_____	45 cm = _____ m
millimeter? _____	_____	45 mm = _____ m

10

©2000 by Key Curriculum Press
Do not duplicate without permission.

Fill in the missing numbers.

___10___ mm = _____ cm

     ___10___ cm = _____ dm

          _____ dm = ___1___ m

___1000___ mm = _____ cm =   _____ m

          _____ m = ___1___ dam

     ___100___ m = _____ hm

     _____ m = ___1___ km

Use the chart on page 10 to change each length below to meters. Here are the symbols for three of the measurements you just learned:
hm = hectometer, dm = decimeter, and dam = decameter.

4 km = _____ m   13 dm = _____ m   445 mm = _____ m

125 cm = _____ m   57 km = _____ m   2347 cm = _____ m

75 hm = _____ m   467 dam = _____ m   1573 mm = _____ m

67 km = _____ m   23 dm = _____ m   546 cm = _____ m

8455 cm = _____ m   47 hm = _____ m   886 dm = _____ m

©2000 by Key Curriculum Press
Do not duplicate without permission.

Planetary Jumping

The gravity of Earth is different from the gravity of other planets. Suppose the highest you could jump on Earth was 1 meter. On Jupiter, you could jump only about 38 centimeters because the gravity there is greater than on Earth.

Complete the chart below and answer the questions at the bottom of the page.

Planet	Height of your jump	Height of your jump in meters
Mercury	278 cm	
Venus	111 cm	
Earth	100 cm	1 m
Mars	26.3 dm	
Jupiter	37.7 cm	
Saturn	877 mm	
Uranus	930 mm	
Neptune	741 mm	
Pluto	4350 mm	

Which has greater gravity, Venus or Neptune? _____

Which planet has the most gravity? _____

Which planet has the least gravity? _____

On which planets could you jump higher than you can on Earth?

_____

©2000 by Key Curriculum Press
Do not duplicate without permission.

Review: Adding, Subtracting, and Multiplying Decimals

Add and subtract. Be sure to line up the decimal points.

$$\begin{array}{r} 2.7 \text{ cm} \\ +4.5 \text{ cm} \\ \hline \end{array} \qquad \begin{array}{r} 7.7 \\ -2.8 \\ \hline \end{array} \qquad \begin{array}{r} 10.06 \text{ m} \\ +3.7 \text{ m} \\ \hline \end{array}$$

$24.3 + 18.8 =$ $\qquad$ $87.57 - 9.3 =$ $\qquad$ $15.7 - 1.95 =$

$234 + 84.3 =$ $\qquad$ $16.7 + 8 =$ $\qquad$ $8 - 0.2 =$

Multiply. The total number of decimal digits in the factors equals the number of decimal digits in the product.

$$\begin{array}{r} 0.97 \\ \times\ 0.3 \\ \hline 0.291 \end{array} \quad \begin{array}{l} \text{2 decimal points} \\ \text{1 decimal point} \\ \text{3 decimal points} \end{array} \qquad \begin{array}{r} 0.9 \\ \times\ 0.8 \\ \hline \end{array} \qquad \begin{array}{r} 2.73 \\ \times\ 5 \\ \hline \end{array} \qquad \begin{array}{r} 0.15 \\ \times\ 0.7 \\ \hline \end{array}$$

$2.37 \times 6 =$ $\qquad$ $2.37 \times 0.6 =$ $\qquad$ $3.6 \times 1.5 =$

$0.7 \times 9 =$ $\qquad$ $5.04 \times 4 =$ $\qquad$ $9.2 \times 8.3 =$

©2000 by Key Curriculum Press
Do not duplicate without permission.

Olympic Records

Some Women's Olympic High-Jump Winners		Some Men's Olympic Long-Jump Winners	
1932 Jean Shiley, USA	1.66 m	1920 William Petterssen, Sweden	7.15 m
1936 Ibolya Csák, Hungary	1.60 m	1936 Jesse Owens, USA	8.06 m
1960 Iolanda Balas, Romania	1.85 m	1968 Robert Beamon, USA	8.90 m
1980 Sara Simeoni, Italy	1.97 m	1980 Lutz Dombrowski, East Germany	8.54 m
1988 Louise Ritter, USA	2.03 m	1984 Carl Lewis, USA	8.54 m
1992 Heike Henkel, Germany	2.02 m	1988 Carl Lewis, USA	8.72 m

Use the information in the chart above to fill in each blank.

• Sara Simeoni jumped _____ higher than Iolanda Balas.

• Robert Beamon jumped _____ farther than Jesse Owens.

• What is the difference between the women's high-jump heights in 1932 and 1992? _____

• In general, the men's Olympic long-jump distances are approximately _____ times as long as the women's high-jump distances.

• How much farther did Carl Lewis jump in 1988 than in 1984? _____

• The difference between the longest and shortest men's long jumps is _____.

• What is the difference between the highest and lowest women's high jumps? _____

©2000 by Key Curriculum Press
Do not duplicate without permission.

Review: Measuring in Centimeters and Millimeters

Measure each line segment to the nearest millimeter. Use your own ruler or tape #2 on page 45.

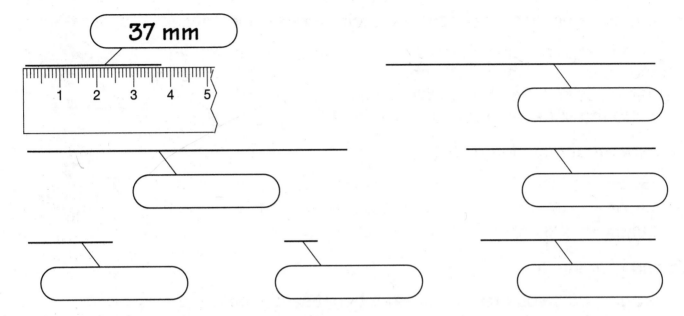

37 mm

Measure each line segment to the nearest tenth of a centimeter.

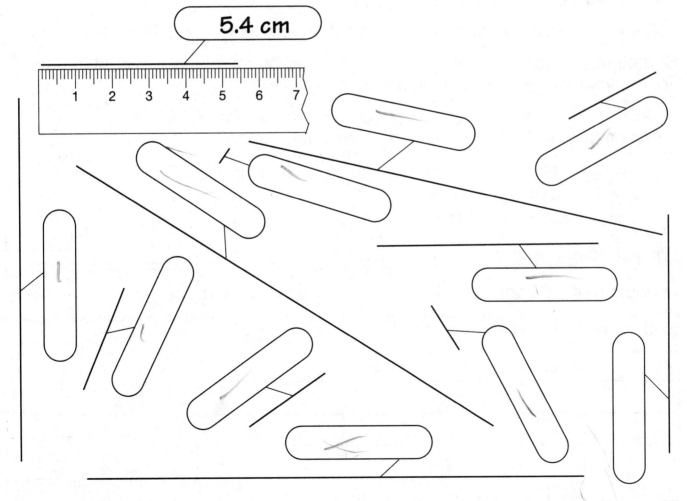

5.4 cm

©2000 by Key Curriculum Press
Do not duplicate without permission.

Renting Yourself Some Ski Equipment

At many ski rental shops, ski boots and skis are measured in centimeters. To measure your proper ski length, you need a small piece of masking tape and a measuring tape.

Here is one way people have found their proper ski length:

Stand next to a wall, extend your arm above your head, and bend your wrist. Record the height of your fingers by placing a small piece of tape on the wall. Measure the height from the floor to the spot you marked in centimeters. _____ cm

Below are the standard lengths of skis.

150 cm	160 cm	170 cm	175 cm	180 cm
185 cm	190 cm	195 cm	200 cm	205 cm
210 cm	215 cm	220 cm	225 cm	

To find your length:

Round up if you are a good skier because it's harder to ski on longer skis.

Round down if you are a beginning skier.

I should rent skis that are _____ centimeters long.

Sometimes shoe and boot sizes are indicated by their length in centimeters. To find your proper ski boot length:

Take off your shoe and measure your foot to the nearest tenth of a centimeter. Ski boots come in the following lengths:

15 cm	16 cm	17 cm	18 cm	19 cm	20 cm	21 cm
22 cm	23 cm	24 cm	25 cm	26 cm	27 cm	28 cm
29 cm	30 cm	31 cm	32 cm	33 cm	34 cm	

Will you round up or down for ski boots? _____ Why? _____

I should rent ski boots that are _____ centimeters long.

Find the best ski and ski boot length for three other students in your class.

Name	Ski length	Ski boot size

©2000 by Key Curriculum Press
Do not duplicate without permission.

Metric Tools

Use a ruler with millimeters or tape #2 on page 45 to complete this page.

A wrench is labeled in millimeters by the size of its opening. Measure each opening.

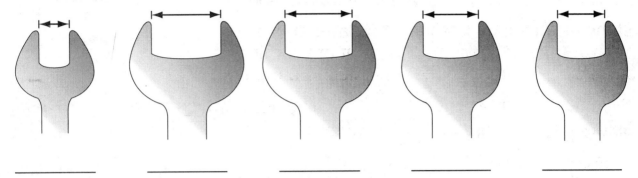

_____   _____   _____   _____   _____

A metric nut is measured in millimeters by the diameter of its hole. Measure each hole size.

_____   _____   _____   _____   _____

A metric bolt is measured in millimeters by its length and its diameter. Measure the length and diameter of each bolt.

length _____ diameter _____

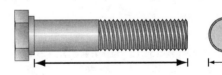

length _____ diameter _____

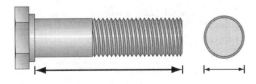

length _____ diameter _____

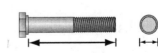

length _____ diameter _____

©2000 by Key Curriculum Press
Do not duplicate without permission.

Olympic Training Center—Using Meters and Kilometers

Olympic-size swimming pool

Swimming

An Olympic-size swimming pool is 50 meters long. Fill in the blanks.

50 m

An athlete who completes the 1500-meter

freestyle swims _____ pool lengths.

A swimmer won an event in the 1996 Olympics in which she swam 16 lengths. How many meters did she swim? _____

A team of four swims the 800-meter freestyle relay. If each swimmer

on the team swims an equal distance, how many lengths

does each swim? _____

Running

- Which event is actually 3 kilometers?

  _____

- Which event is closest to 0.5 kilometer?

  _____

- Which event is 1.5 kilometers?

  _____

- Which event is 5 kilometers?

  _____

- In the 4 × 400-meter relay, four athletes
  each run 400 meters. Therefore, the
  4 × 400-meter relay is _____ kilometer(s)
  and _____ meter(s).

Some Olympic events
100-meter run
200-meter run
400-meter run
800-meter run
1500-meter run
4 × 100-meter relay
4 × 400-meter relay
3000-meter steeplechase
5000-meter run
10 000-meter run

©2000 by Key Curriculum Press
Do not duplicate without permission.

More About Kilometers

One kilometer is about the length of 11 football fields.
1 kilometer = 1000 meters

Walking 1 kilometer takes about 15 minutes.	Jogging 1 kilometer takes about 7 minutes.	Biking 1 kilometer takes about 4 minutes.

Is your home less than 1 kilometer, about 1 kilometer, or greater than
1 kilometer from the closest park? _____

Is your home less than 1 kilometer, about 1 kilometer, or greater than
1 kilometer from the closest grocery store? _____

It takes Hilda 30 minutes to walk from her home to school. About how far
does she walk? _____

Tim biked 6 kilometers. About how long do you think it took him?

_____

George walked to and from a park 1.5 kilometers away. About how long do
you think the entire trip took him? _____

The distance from Mei Mei's house to Susan's house is half a kilometer.
About how long would it take to jog there? _____

Circle the best estimate.

length of a jump rope
2 mm   2 m   2 km

length of Lake Huron
329 cm   329 m   329 km

length of a mail carrier's route
10 cm   10 m   10 km

length of an alligator
3 cm   3 m   3km

length of a large insect
19 mm   19 cm   19 km

height of the Empire State Building
381 cm   381 m   381 km

greatest depth of the Pacific Ocean
11 cm   11 m   11 km

air distance from San Francisco to
New York City
4137 cm   4137 m   4137 km

length of a car
4 cm   4 m   4 km

driving distance from Chicago, Illinois,
to Boise, Idaho
2256 cm   2256 m   2256 km

©2000 by Key Curriculum Press
Do not duplicate without permission.

A Distance Chart for Mexican Cities

The chart below shows driving distances in kilometers between some cities in Mexico.

	Mérida	Chihuahua	Guadalajara	Mexico City	Monterrey
Mérida	0	2816	1960	1407	2142
Chihuahua	2816	0	1138	1410	819
Guadalajara	1960	1138	0	458	737
Mexico City	1407	1410	458	0	924
Monterrey	2142	819	737	737	0

Use this distance chart to answer each question.

It is ___458___ kilometers from Guadalajara to Mexico City.

How far is it from Chihuahua to Monterrey? _____

How far is it from Mérida to Guadalajara? _____

How far is it between each pair of cities?

Monterrey and Guadalajara _____

Monterrey and Mérida _____

Which distance is farther?

_____

You plan to drive from Mérida to Monterrey
but you want to stop in Mexico City on your way.

How far do you drive from Mérida to Mexico City? _____

How far do you drive from Mexico City to Monterrey? _____

What is the total distance you drive? _____

How much farther is your trip than it would have been if you had gone

directly from Mérida to Monterrey? _____

José is driving from Chihuahua to Mexico City. He has already driven

989 kilometers. How many more kilometers does he have to drive? _____

20

©2000 by Key Curriculum Press
Do not duplicate without permission.

Odometer Readings

A car odometer measures how far a car has gone since it was made. In countries other than the United States, the odometer measures kilometers. The pictures below show how an odometer changes as the car travels.

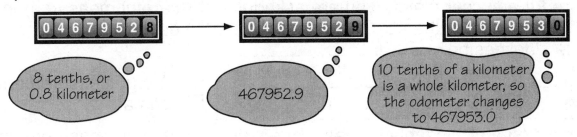

8 tenths, or 0.8 kilometer

467952.9

10 tenths of a kilometer is a whole kilometer, so the odometer changes to 467953.0

Write the next two odometer readings, going up in tenths of a kilometer.

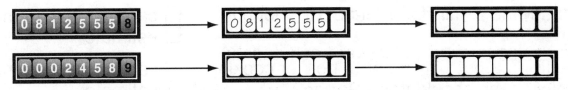

Here's how Christina used the odometer on her mother's car to find the distance from home to basketball practice.

Step 1 She recorded the starting odometer reading at home.

0 0 7 9 3 2 5

Step 2 She recorded the ending odometer reading when she got to practice.

0 0 7 9 3 4 8

Step 3 Then she subtracted the starting reading from the ending reading.

7934.8
−7932.5

Step 4 The distance from her house to basketball practice is 2.3 kilometers, or 2300 meters.

Subtract to find each distance. Then record the distance in kilometers and in meters.

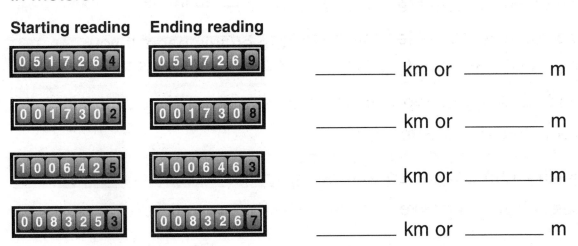

Starting reading   **Ending reading**

0 5 1 7 2 6 4   0 5 1 7 2 6 9   _____ km or _____ m

0 0 1 7 3 0 2   0 0 1 7 3 0 8   _____ km or _____ m

1 0 0 6 4 2 5   1 0 0 6 4 6 3   _____ km or _____ m

0 0 8 3 2 5 3   0 0 8 3 2 6 7   _____ km or _____ m

©2000 by Key Curriculum Press
Do not duplicate without permission.

Estimation or Careful Measurement?

Decide for each situation if you need to take a careful measurement or if a rough estimate is good enough. Mark the box that shows your answer.

To prepare for a soccer match, you are asked to mark the boundaries of the field. How long should the sidelines be?
☐ careful measurement
☐ estimate

You are decorating the gymnasium with streamers. How much material should you buy?
☐ careful measurement
☐ estimate

You are building a picnic table to match one you already own. How long should the new table be?
☐ careful measurement
☐ estimate

You are sewing a shirt for a friend. How long is her arm?
☐ careful measurement
☐ estimate

You are filling a pot with water to boil for cooking pasta. How deep should the water be?
☐ careful measurement
☐ estimate

You are giving directions to a friend. How many kilometers should you tell her to drive down a certain street before turning right?
☐ careful measurement
☐ estimate

The recipe says you should cut the flat cake into 5-centimeter squares to get 16 servings. Where should the cuts be?
☐ careful measurement
☐ estimate

You are mapping out a route for a 10-kilometer foot race. Where should the starting line and finish line be?
☐ careful measurement
☐ estimate

You want to buy a desk to fit into a small space. How wide can the desk be?
☐ careful measurement
☐ estimate

You want to tie three gifts with fancy ribbon. How much ribbon should you buy?
☐ careful measurement
☐ estimate

It is possible to disagree on some of the answers on this page. If you disagree with another classmate or the answer book, explain your reasoning here.

_____

_____

_____

©2000 by Key Curriculum Press
Do not duplicate without permission.

Introduction to Perimeter

The **perimeter** is the distance around a figure. For each figure, measure each side. Use a ruler or one of the tapes on page 45. Then find the perimeter by adding the lengths of the sides. Record your answer to the nearest centimeter.

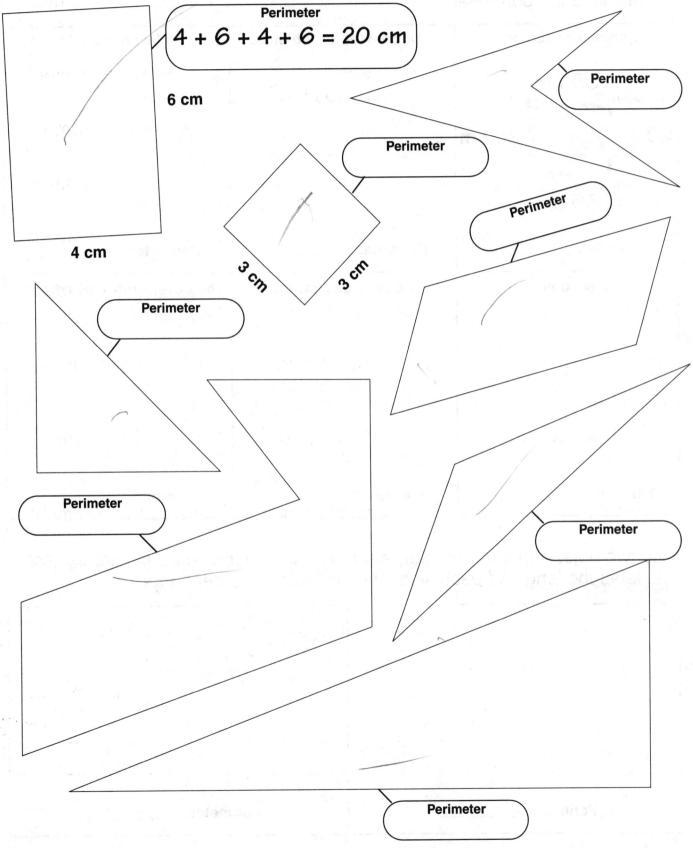

Perimeter

$4 + 6 + 4 + 6 = 20$ cm

6 cm

4 cm

3 cm

3 cm

Perimeter

Perimeter

Perimeter

Perimeter

Perimeter

Perimeter

Perimeter

©2000 by Key Curriculum Press
Do not duplicate without permission.

Measuring Perimeters

- Draw a picture of each object mentioned below.
- Measure and label the length of each side to the nearest centimeter.
- Calculate the perimeter.

a computer screen:	your desktop or tabletop:	a favorite book:
35 cm 28 cm 28 cm 35 cm		
Perimeter: _____	Perimeter: _____	Perimeter: _____
a window:	**a poster or picture:**	**the cover of this workbook:**
Perimeter: _____	Perimeter: _____	Perimeter: _____

Find two objects in the room that are *not* rectangular. Draw a picture of each and label the length of each side. Then calculate the perimeter.

Perimeter: _____	Perimeter: _____

©2000 by Key Curriculum Press
Do not duplicate without permission.

Exploration of Perimeters

Draw four shapes that are *not* rectangles below, each with a perimeter of 10 centimeters.

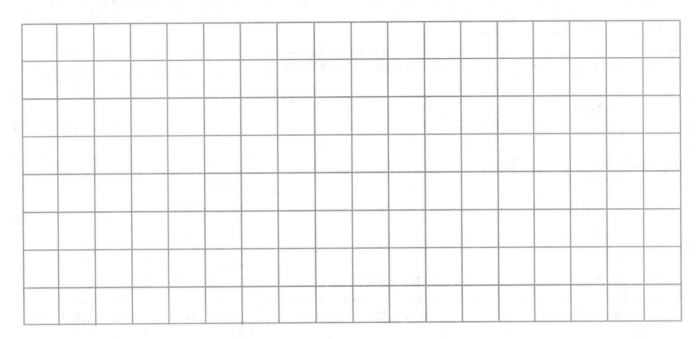

Does each shape have the same number of squares inside?   ☐ Yes   ☐ No

Draw two rectangles that have the same perimeter
but different lengths.

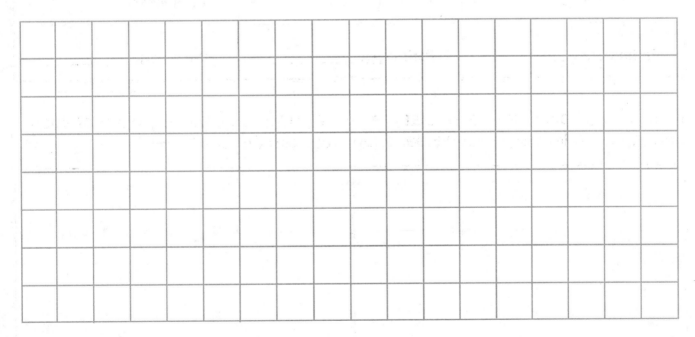

The perimeter of each of my rectangles is _____ centimeters. Does each
rectangle have the same number of squares inside?   ☐ Yes   ☐ No

©2000 by Key Curriculum Press
Do not duplicate without permission.

Calculating Perimeters

If two sides of the same shape are marked with the same number of tick marks, the sides are the same length. Find the perimeter of each shape using only the measurements and marks shown. (You do not need your ruler.)

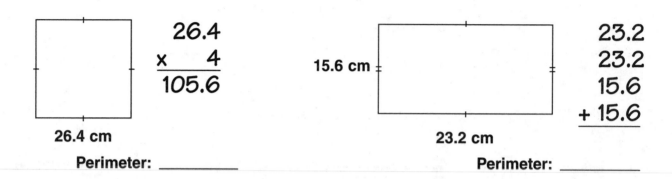

$$26.4 \times 4 = 105.6$$

26.4 cm

Perimeter: _____

15.6 cm

23.2 cm

$$23.2$$
$$23.2$$
$$15.6$$
$$+ 15.6$$

Perimeter: _____

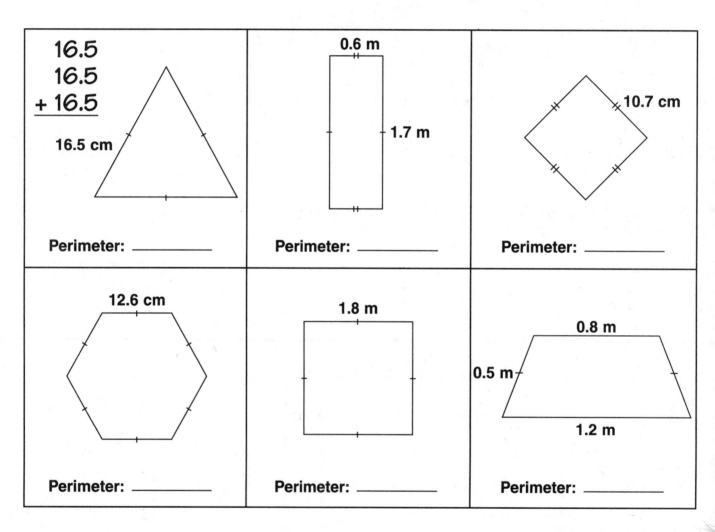

$$16.5$$
$$16.5$$
$$+ 16.5$$

16.5 cm

Perimeter: _____

0.6 m

1.7 m

Perimeter: _____

10.7 cm

Perimeter: _____

12.6 cm

Perimeter: _____

1.8 m

Perimeter: _____

0.8 m

0.5 m

1.2 m

Perimeter: _____

©2000 by Key Curriculum Press
Do not duplicate without permission.

Missing Measurements

Here is an irregular figure with the measurement of one side missing.

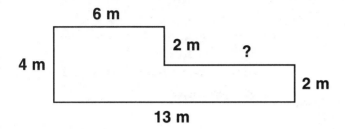

To find its perimeter, first you must find the missing length.

Step 1 Find the missing length.

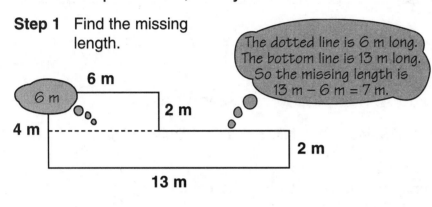

The dotted line is 6 m long. The bottom line is 13 m long. So the missing length is 13 m − 6 m = 7 m.

Step 2 Find the sum of the lengths of the sides.

13
4
6
2
7   The
+ 2   perimeter
34   is 34 m.

Find each missing length. Then find the perimeter of each figure.

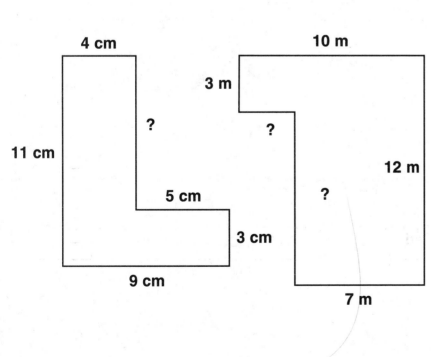

Perimeter: _____

Perimeter: _____

Perimeter: _____

©2000 by Key Curriculum Press
Do not duplicate without permission.

Team Project #2: The Secret of Circumference

You will discover a special relationship between diameter and circumference on the next few pages.

The longest distance, *d*, across a circle is called the diameter of the circle. The diameter goes through the circle's center.

c

d

The distance, *C*, around a circle is called the circumference of the circle.

Center of the circle.

On the circle above, what does *C* stand for? _____

What does *d* stand for? _____

For each circle on this page and the next, follow the steps below.

Step 1 Use a tape measure or a piece of string to carefully measure the circumference to the nearest centimeter.

Step 2 Draw the diameter. Use a ruler to measure it to the nearest centimeter.

Step 3 Use a calculator to find the *length of the circumference* divided by the *length of the diameter, C ÷ d.* Round your answer to the nearest whole number.

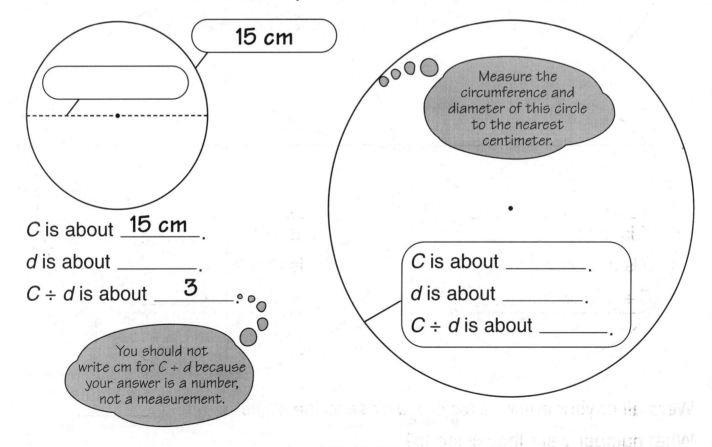

15 cm

C is about ___15 cm___.

d is about _____.

C ÷ *d* is about ___3___.

You should not write cm for *C* ÷ *d* because your answer is a number, not a measurement.

Measure the circumference and diameter of this circle to the nearest centimeter.

C is about _____.

d is about _____.

C ÷ *d* is about _____.

©2000 by Key Curriculum Press
Do not duplicate without permission.

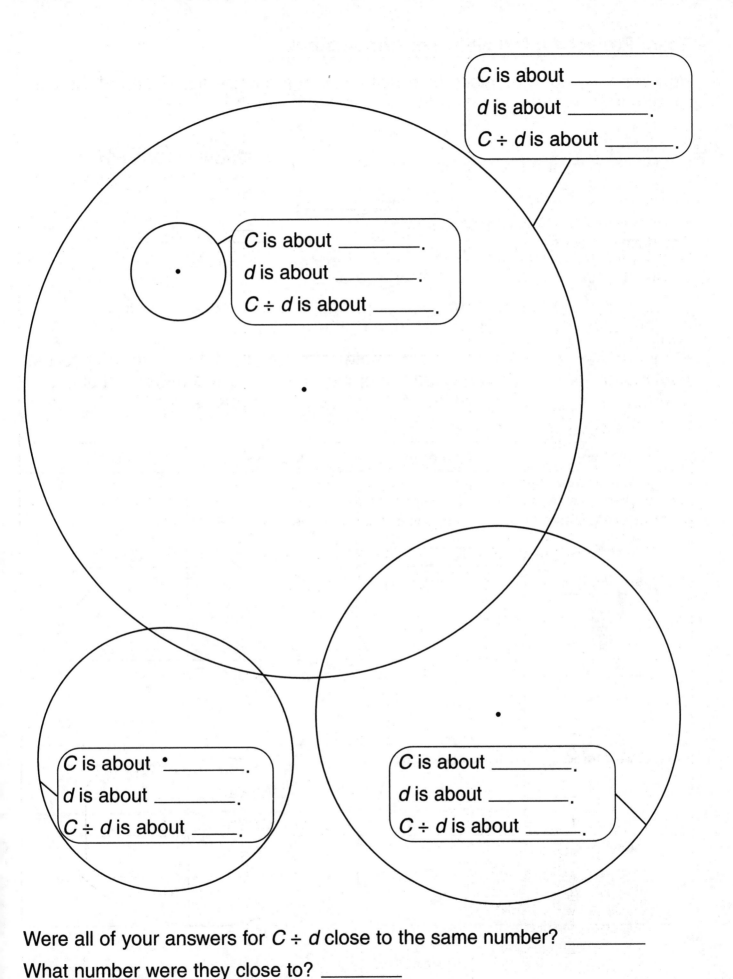

Were all of your answers for $C \div d$ close to the same number? _____

What number were they close to? _____

©2000 by Key Curriculum Press
Do not duplicate without permission.

29

Team Project #3: Circumference All Around

You need a can and a piece of paper. Look at the circle at the end of the can. In this project, you'll measure the circumference of that circle.

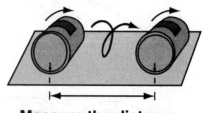

- Make a mark on the rim of your can for the starting point.

- Mark a starting point on your paper.

- Match up the two starting points.

Measure the distance.

- Roll the can around one full turn until the starting point touches the paper again.

- Measure on the paper from the starting point to the ending point. This distance is the circumference of the circle.

Measure the circumference and diameter of each object below to the nearest centimeter. Use a measuring tape, a string, or the method described above. You may use a calculator to find $C \div d$. Round your answer for $C \div d$ to the nearest tenth.

Circular object	Circumference	Diameter	$C \div d$
small child's bike tire			
small can			
your choice			
your choice			
your choice			

What conclusion can you make about $C \div d$? _____

©2000 by Key Curriculum Press
Do not duplicate without permission.

Did you discover that $C \div d$ is about 3.1 for every circle you tried?   ☐ Yes ☐ No

Mathematicians have a special name for $C \div d$ because it is the same for every circle.

$C \div d$ is called **pi** and is written using the Greek symbol π. You pronounce it like the word "pie."

Write π three times: _____ _____ _____

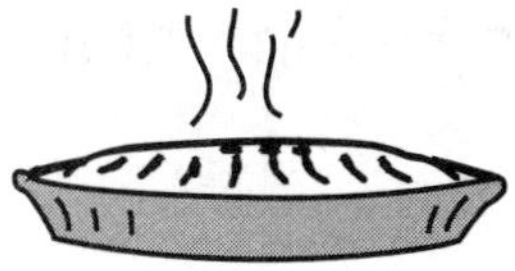

On many calculators there is a π key. If you press it, you will see the first several digits of the actual value of π. The 3 is the whole-number part, and the rest is the decimal part.

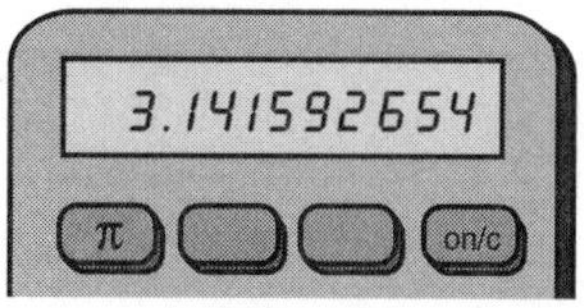

3.141592654
π ☐ ☐ on/c

Check the correct choice in each sentence below.

$C \div d$ is a little more than ☐ 3 ☐ 4.

The circumference of a circle is ☐ longer   than the diameter.
                                  ☐ shorter

For each pair, check the sentence that makes more sense.

☐ The circumference is about 3 times the diameter.
☐ The circumference is about one-third of the diameter.

☐ The diameter is about 3 times the circumference.
☐ The diameter is about one-third of the circumference.

Pi (π) written to the nearest tenth is 3.1. Use this estimate of pi to complete the table.

Object	Circumference	Diameter
soup can top	18 cm	6 cm
dinner plate	75 cm	
large pizza		45 cm
Earth		12750 km
face of a round watch		3 cm

©2000 by Key Curriculum Press
Do not duplicate without permission.

Can You Read a Floorplan?

The diagram on the right is called a **floorplan.** Architects, home decorators, and builders use floorplans to show what a room or a structure would look like if the roof were removed and you were looking at it from above.

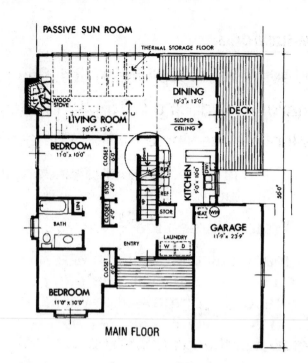

Here is a floorplan of George's bedroom. Use this floorplan to answer the questions below.

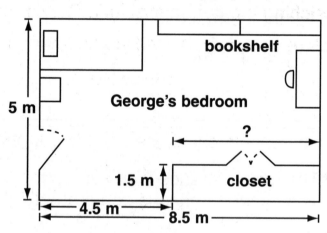

What is the length of George's room?

_____

What is the width of George's room?

_____

What is the length of George's closet?

_____

What is the width of the closet?

_____

What is the perimeter of George's room?

_____

What is the perimeter of George's closet?

_____

©2000 by Key Curriculum Press
Do not duplicate without permission.

Reductions

On some photocopy machines, you can make a picture or a photograph smaller by reducing it. The horse pictured in the photocopy below is _____ as long as the horse in the original picture.

original picture

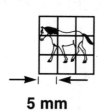

1 cm ⟶⊢   ⊢◄—

reduced photocopy

5 mm

Below is an original picture of a cat and part of a reduced photocopy. Finish sketching the cat on the right. The cat is _____ times as long in the original picture as it is in the reduced photocopy.

original picture

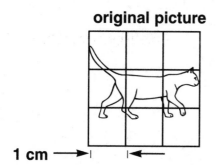

1 cm ⟶⊢   ⊢◄—

reduced photocopy

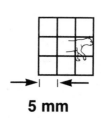

→⊢ ⊢◄

5 mm

Each length in the original picture below is twice as long as it will be in the reduced photocopy. Sketch a reduction of the squirrel in the smaller square. The squirrel is _____ tall in the original picture. How tall will it be in the reduced picture? _____ The squirrel's tail is _____ long in the original picture. How long will it be in the reduced picture? _____

original picture

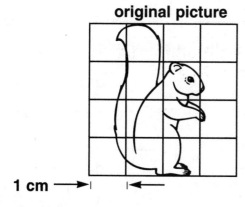

1 cm ⟶⊢   ⊢◄—

reduced picture

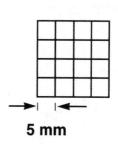

→⊢ ⊢◄—

5 mm

©2000 by Key Curriculum Press
Do not duplicate without permission.

Make a reduced picture of your desktop or tabletop.

- Measure the length and width of the top to the nearest decimeter. Copy the decimeter strip on this page to help you measure, if needed.

- Draw the reduced picture on the grid below. Make each decimeter on your top equal to 1 centimeter on the grid.

- Include in your drawing some of the objects on your desk.

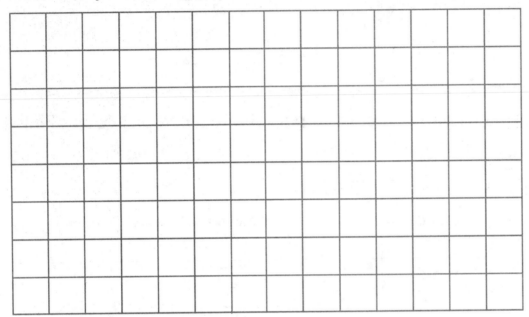

Follow the same procedure as above to draw reduced pictures of two or three other objects in the room. You may view an object from above or from the side.

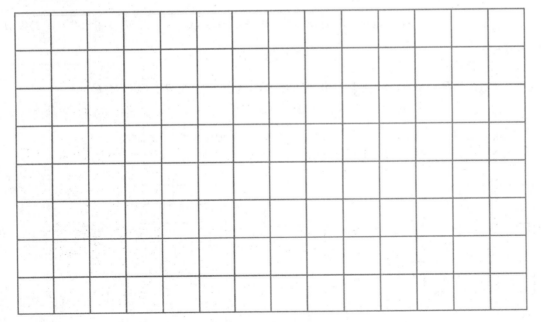

1 decimeter

©2000 by Key Curriculum Press
Do not duplicate without permission

Draw a floorplan of the room you are in. Show all walls and doors. Measure everything to the nearest meter. Make every meter you measure equal to 1 centimeter on the picture below. If your room is small, you may have space to show another room or hallway.

©2000 by Key Curriculum Press
Do not duplicate without permission.

Scale and Scale Drawings

The floorplan you drew on page 35 is an example of a **scale drawing.** A map of a park or a country is also a scale drawing. A scale drawing is a picture with the same shape as the original but a different size. Every scale drawing needs a **scale** to show the actual distance on the original and the corresponding distance on the drawing.

Match each description with a scale drawing below.

scale drawing of a house	scale drawing of the state of Arizona	scale drawing of a bolt

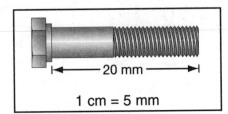

1 cm = 5 mm

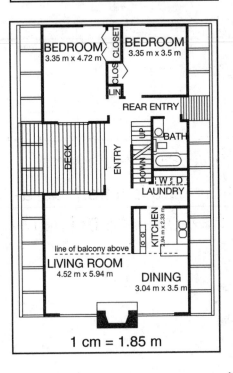

1 cm = 1.85 m

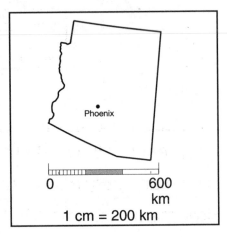

1 cm = 200 km

The scale on the map at right says that 500 kilometers in the real world is represented by 1-centimeter on the map.

Use a ruler or a tape to measure the distance on the map between Halifax and Montreal. It is about _____ centimeters.

If you actually flew the distance from Halifax to Montreal, it would be about _____ kilometers.

©2000 by Key Curriculum Press
Do not duplicate without permission.

The diagram below shows a student's scale drawing of her classroom.

One centimeter on the diagram below represents _____ in the real classroom.

On the diagram, how many centimeters is it from the doorway to the round table? _____

How many meters is it from the real doorway to the real round table? _____

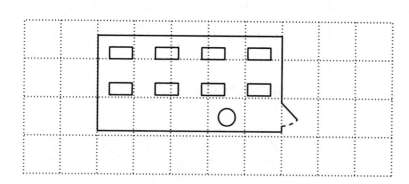

Scale

0   3 m
├────────┤
0   1 cm

One centimeter on the map of Texas represents _____ in the real state.

How many *centimeters* on the map is it from Dallas to El Paso? _____

How many *kilometers* is it from Dallas to El Paso? _____

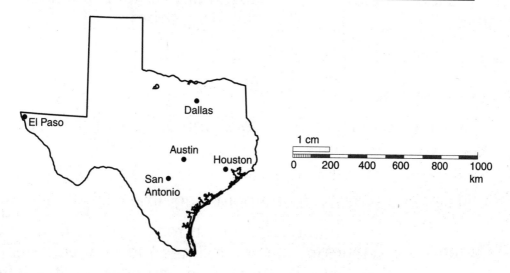

Now use the scale to figure out some more distances.

How many kilometers is it from Dallas to Houston? _____

From San Antonio to Austin? _____

From Austin to Houston? _____

What is the longest straight distance you can travel, staying in Texas the entire time? _____

©2000 by Key Curriculum Press
Do not duplicate without permission.

Scale Drawing of Maria's House

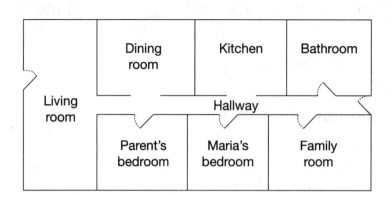

The scale tells you that _____ on the scale drawing represents _____ in the real house.

Use a ruler to measure each distance on the scale drawing to the nearest tenth of a centimeter. Find the real length in meters when necessary.

In the scale drawing, how many centimeters long is the living room?

_____

In the scale drawing, how many centimeters wide is the living room?

_____

How many meters long is the real living room?

_____

How many meters wide is the real living room?

_____

You want to carpet both bedrooms.

In the scale drawing, how many centimeters long are they together?

_____

How many meters long are the real bedrooms together? _____

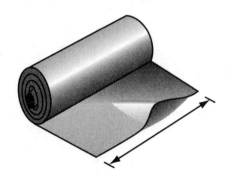

You can buy carpet in rolls that are 4 meters wide. How long will the roll need to be if you want to carpet both bedrooms and have as little leftover carpet as possible? Explain.

_____

_____

©2000 by Key Curriculum Press
Do not duplicate without permission.

Use the scale drawing on page 38 to find the actual lengths. Then write each response as a sentence.

1. the length of the family room in meters

The length of the family room is 5.6 meters.

2. the width of the family room in meters

_____

_____

3. the length of the hallway in meters

_____

_____

4. the width of the hallway in meters

_____

_____

5. the length of the kitchen in meters

_____

_____

6. the width of the kitchen in meters

_____

_____

7. the perimeter of the living room in meters

_____

_____

8. the distance from the front doorway to the back doorway in meters

_____

_____

9. the length of the bathroom and the kitchen together in meters

_____

_____

10. the distance from the family room door to the front door in meters

_____

_____

11. the perimeter of the house in meters

_____

_____

12. the length of Maria's bedroom in meters

_____

_____

©2000 by Key Curriculum Press
Do not duplicate without permission.

Scale on a Map

The map below shows the country of Japan.

On this map, 1 cm represents _____ km.

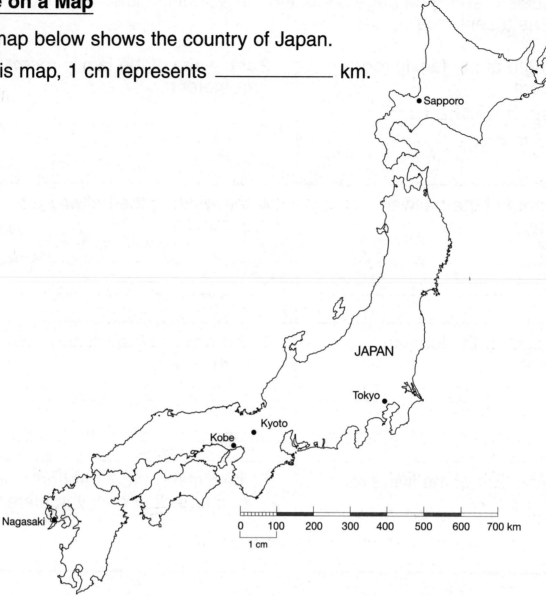

About how many kilometers is it from Kobe to Tokyo?

_____

About how far is it from Nagasaki to Kyoto?

_____

How far is Nagasaki from Sapporo?

_____

What is the distance in kilometers from Tokyo to Sapporo?

_____

How far is it from Tokyo to Nagasaki?

_____

What is the least number of kilometers you would need to travel in order to visit all five cities?

_____

©2000 by Key Curriculum Press
Do not duplicate without permission.

Practice Test

Fill in the blanks.

7 m = _____ cm   13 mm = _____ cm   2 m = _____ mm

400 cm = _____ m   26 cm = _____ mm   64 cm = _____ m

280 mm = _____ cm   5 m = _____ cm   823 cm = _____ m

347 cm = _____ m plus _____ cm   3.57 m = _____ m plus _____

Change the larger unit to the smaller unit. Circle the longer measurement.

16 m   **1500 cm**   **20 cm**   **220 mm**

_____ cm   _____ mm

Fill in the blanks.

68 km = _____ m   1752 mm = _____ m   236 cm = _____ m

14 dm = _____ m   4500 m = _____ km   3257 cm = _____ m

Joe jumped 1.23 meters in the high jump. Terry jumped 1.30 meters. How much higher did Terry jump?

A frog jumped 0.6 meter high. A student jumped 1.2 times higher. How high did the student jump?

_____   _____

Measure to the nearest tenth of a centimeter.

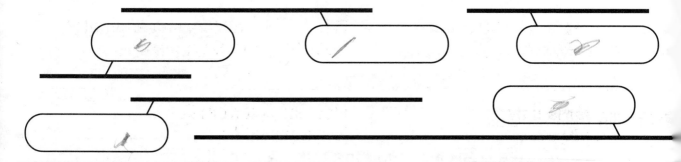

Measure each wrench and metric nut to the nearest millimeter.

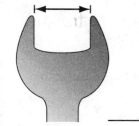   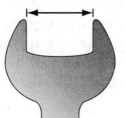   

_____   _____   _____

©2000 by Key Curriculum Press
Do not duplicate without permission.

las	New York City (midtown)	Salt Lake City
550	1340	2355
0	2600	2382

According to the chart, how many kilometers would you travel if you drove from Chicago to Salt Lake City? _____

eadings to find the distance of the car trip in kilometers.

ding    _____

ading

r of each figure. You may need to first find the
ments.

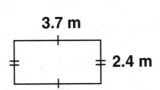

_____   _____

3 m

7 m

4 m

5 m

s. Let π be about 3.14.

ct	Circumference	Diameter
		30 cm
	40.82 cm	

he map, then write the actual distances.

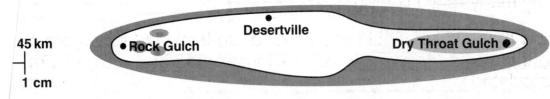

45 km

1 cm

to Rock Gulch _____   Desertville to Dry Throat Gulch _____

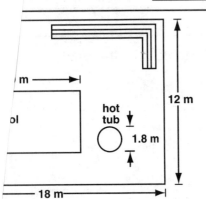

Find the perimeter of the whole pool area shown.

_____

Find the perimeter of the pool. _____

Find the approximate circumference of the hot tub.

©2000 by Key Curriculum Press
Do not duplicate without permission.

Projects

1. Set up a metric Olympics with your classmates. Include the standing and running broad jumps, 100-meter run, and 400-meter relay. Record the jumps in meters or in centimeters. Describe your Olympics and report on your results.

2. The film for most cameras is called "35 mm film." Look at some film of this type. Some part of it measures 35 millimeters. What is it?

3. Look at the inside cover of this book and read "Numbers in Many Languages." Describe any relationships you see in the chart. Next, find the words for the same numbers in another language. Describe how the words compare with those in the chart.

4. Make a scale drawing of your bedroom at home. Show closets, doors, and some furniture. Be sure to show your scale.

5. Make a metric distance chart for your town. If you need to, have a parent or an older sibling who can drive help you figure out each distance below. You can also use a map. To change miles to kilometers, multiply miles by 1.6.

	your house	grocery store	school
your house			
grocery store			
school			

6. Find a map of another part of the world with a scale in kilometers. Plan a trip in that region that stops in at least five places. Calculate how many total kilometers your trip will be.

7. How many times would you have to run back and forth from one end of your block to the other to run 1 kilometer? Explain how you measured and calculated your answer.

©2000 by Key Curriculum Press
Do not duplicate without permission.

Measuring Tools

Use the rulers and the tape measures on this page to do the problems in this book. After each use, tuck them safely in your book in case you need to use them again.

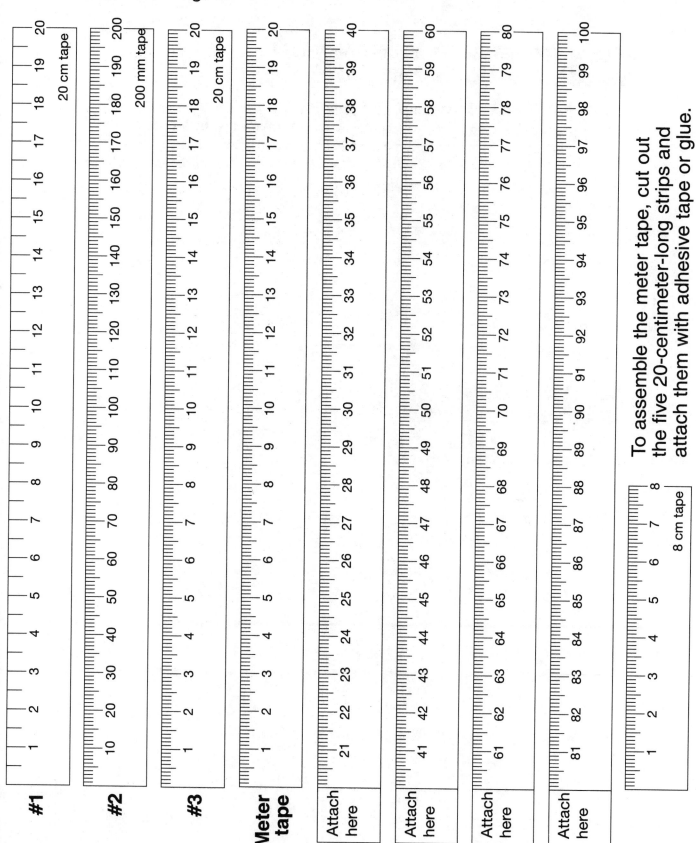

To assemble the meter tape, cut out the five 20-centimeter-long strips and attach them with adhesive tape or glue.

©2000 by Key Curriculum Press
Do not duplicate without permission.

Key to Metric Measurement®

Book 1: *Metric Units of Length*
Book 2: *Measuring Length and Perimeter Using Metric Units*
Book 3: *Finding Area and Volume Using Metric Units*
Book 4: *Metric Units for Mass, Capacity, Temperature, and Time*
Answers and Notes for Books 1–4

Also Available

Key to Fractions®
Key to Decimals®
Key to Percents®
Key to Algebra®
Key to Geometry®
Key to Measurement®

KEY CURRICULUM PRESS
Innovators in Mathematics Education

ISBN 1-55953-326-9

9 781559 533263